Spotter's Guide to
THE NIGHT SKY

Nigel Henbest

Special Consultant
Dr Lloyd Motz

Professor of Astronomy, Columbia University
in the City of New York

Illustrated by Michael Roffe

This is a telescope view of the
Great Nebula in the constellation
of Orion. See where to look for it on
page 19; check it off when you have
seen it (page 27).

MAYFLOWER BOOKS
NEW YORK

Contents

Designed and edited by David Jefferis Ltd
© 1979 by Usborne Publishing Limited
First American edition.

Manufactured in Spain by Printer, industria
gráfica sa Sant Vicenç dels Horts, Barcelona
D.L. 14.141-1979

**Library of Congress Cataloging in
Publication Data**
Henbest, Nigel.
    Spotter's guide to the night sky.
    Includes index.
    1. Astronomy – Observers' manuals. I. Title.
QB63.H38   523   79-605
ISBN 0-8317-6375-2
ISBN 0-8317-6376-0 pbk.

How to use this book

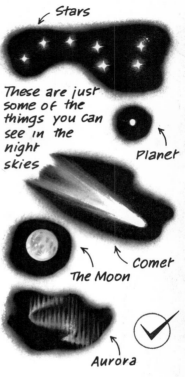

Stars

These are just some of the things you can see in the night skies

Planet

Comet

The Moon

Aurora

This book is an identification guide to the wide range of things you can see in the sky at night. Take it with you when you go out observing on a clear night. Not all the objects in this book will be visible on any particular night, but over the course of a year you should be able to see most of them.

Some objects, like the stars, are very distant, while the planets, comets and the Moon are much closer, though even the Moon is a long way away – over 236,000 miles. A few of the "lights" in the night sky, like the aurorae, occur in the Earth's atmosphere. This book starts with distant sky sights and then moves to ones closer to Earth with descriptions to help you identify them.

Read through the book inside the house before you go out, so you know roughly the kinds of things to look out for – how to recognize a satellite, for example. Find out which constellations are visible, using the sky map on page 8. Look to see where the planets will be, and if there will be any meteor showers.

Next to most of the things in the book is a small blank circle. Each time you see an object, check it off in the circle. Some things, like the more distant planets, can only

be seen with powerful telescopes so these do not have "check circles" next to them.

Scorecard

At the end of the book is a score-card that gives you a score for each object you see. A common or easily recognized object scores 5 points; a very rare or faint one is worth up to 50 points. You can add up your score after a night's observation, or whenever you like.

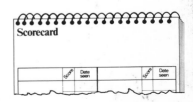

Scorecard

	Score	Date seen		Score	Date seen

Observing the skies

When you go out sky-watching remember to dress warmly. Even in summer you will get cold quickly when sitting still, and in winter you will need to wear two pairs of socks, two sweaters, a warm hat and gloves. Some astronomers even wear two pairs of pants, one over the other.

Find a comfortable garden chair or lounge chair. Standing up soon becomes uncomfortable, and if you lie on the ground you may end up wet from dew or frost. A mat will keep your feet warmer if they touch the ground when you are sitting. Choose a spot in your garden where your sky-view is not blocked by trees, and, if you can, keep away from street lights.

When you come out of a brightly-lit house into a dark night, your eyes will take about half an hour to adjust to the dark. At first you will see just the brighter stars, so wait a little before searching for the fainter objects. Use a red light or a very dim flashlight to see this book while you are observing, as powerful light will ruin your eyes' dark adaptation.

Don't have a hot drink before you go out, because it will, surprisingly, soon make you feel cold. Have one when you come in to warm up, especially if you are going straight to bed.

Hat

Socks

Thick pants

Coat

Thick shoes

Two pullovers

Gloves

Flashlight covered with red cellophane

Binoculars and telescopes

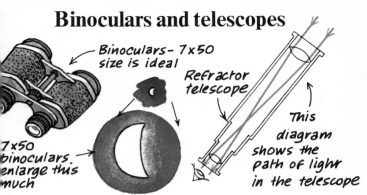

Binoculars– 7 x 50 size is ideal

7 x 50 binoculars. enlarge this much

Refractor telescope

This diagram shows the path of light in the telescope

You can see most of the sky sights in this book with your unaided eyes, but a pair of binoculars will show you much more.

Binoculars are just a pair of telescopes, one for each eye. Telescopes magnify objects, and so they show you more detail than the eye alone can see. A reasonable pair of binoculars (7 x 50 are ideal) will reveal craters on the Moon, and the round globe of Jupiter.

Unfortunately binoculars will also magnify the shaking of your hands as you hold them, and the image will wobble about.

The large front lenses of binoculars gather much more light than your eyes, so the view through binoculars is very bright. Some stars are quite dazzling, and the binoculars will show you many stars that are too faint for your unaided eyes to see.

A telescope is more powerful than binoculars, but more expensive, and many of the cheaper ones are of rather poor quality. Generally it is better to buy a pair of binoculars than a cheap telescope at the same price, if you have the choice. The pictures above and below show the two types of telescope you could use. The refractor uses glass lenses to refract (or bend) the light. The reflector uses a mirror for the same purpose.

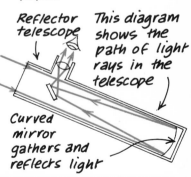

Reflector telescope

This diagram shows the path of light rays in the telescope

Curved mirror gathers and reflects light

WARNING Never, ever, look at the Sun either directly or through binoculars or telescope – you could easily blind yourself.

Our place in the Universe

North Pole   Day

Night   Sun's rays

The Earth rotates once every 23 hours 56 minutes

The Earth is one of nine planets that go around the Sun in circular paths called orbits. You can see above an artist's impression of the Solar System: the name for the Sun, planets, comets and asteroids.

The Earth turns around once a day and completes one revolution about the Sun in a year. Keeping the Earth company is the Moon, a much smaller body which circles the Earth once every 27 days. It is the Earth's only natural satellite, although in the past 20 years hundreds of artificial satellites have been launched by rocket to circle the Earth. Most of the other planets have natural satellites, and Jupiter holds the record with fourteen of them.

Although the planets and the Moon shine brightly in the sky, they are only reflecting light from the Sun; they do not produce their own light. The Sun and planets are shown to the same scale at the top of the page.

6

Asteroids
Mars
Earth   Jupiter   Saturn   Uranus Neptune   Pluto

The picture above shows the Milky Way galaxy, a vast spiral of stars and dust of which the Solar System is a member. Astronomers think that there are about 100,000 million stars in the galaxy. The red arrow shows the position of the Solar System, though on this scale, the Sun is too tiny to be seen.

Distances between stars are vast and are measured in light years, the distance that light travels in a year. Light speed is just over 186,000 miles a second, so a light year (LY) is about 6¼ million million miles. The nearest star, Proxima Centauri, is 4.3 LY away, while the galaxy is 100,000 LY across.

Beyond our galaxy, which you can see in the night sky as a faint band of light called the Milky Way, are billions of others extending into the furthest depths of space. The nearest are the two Magellanic Clouds, visible in the southern hemisphere. They are between 170,000 and 200,000 LY away.

How it all began: the Big Bang

Astronomers think the Universe began in an enormous explosion, called the Big Bang, which happened about 15,000 million years ago. Gas clouds thrown out turned into galaxies, and even today all the galaxies are racing apart from each other as a result of this initial explosion.

Stars of the northern skies

On a clear night you can see about 3,000 stars scattered across the sky. Astronomers find their way around by grouping stars together into patterns, like dot-to-dot puzzles. These 88 patterns, called constellations, are always known by their Latin names; most were first named thousands of years ago.

During the night, the sky seems to rotate, carrying the constellations slowly from east to west. In fact, it is the Earth which is turning, causing some constellations to rise and others to set.

The constellations visible change during the year, as the Earth moves around the Sun and, since the Earth is a globe, people in the northern hemisphere cannot see stars grouped over the South Pole and vice versa.

The sky map on the right shows the brightest stars visible from the northern hemisphere. The stars to be seen from the southern hemisphere are on page 10.

How to use the star map

Find the month in the map margin; turn the book around until the present month is lowest. Sitting in your chair, face south and look for the stars as they appear on the map. You will be able to see most of the stars shown in the center and lower part of the map.

Spot the prominent constellations, then turn to the page numbers marked to find the fainter constellations which are not marked here. The broken lines show the areas covered by each double-page constellation map.

The two small pictures show how the view changes over the year. The

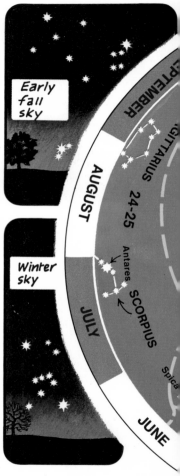

Early fall sky

Winter sky

stars to be seen at one place in the fall, for example, are completely different from those seen from the same place in the winter.

8

Signposts in the sky

Extend the imaginary lines joining the stars in the directions shown by the yellow arrows, to pinpoint bright stars and other constellations.

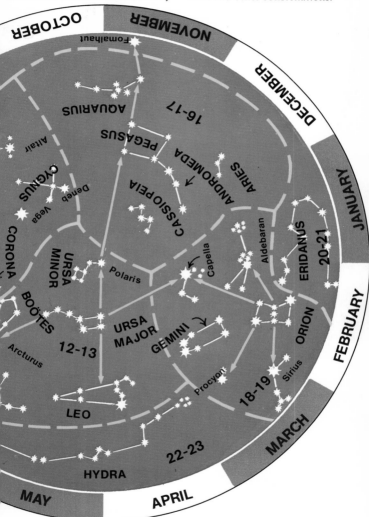

9

Stars of the southern skies

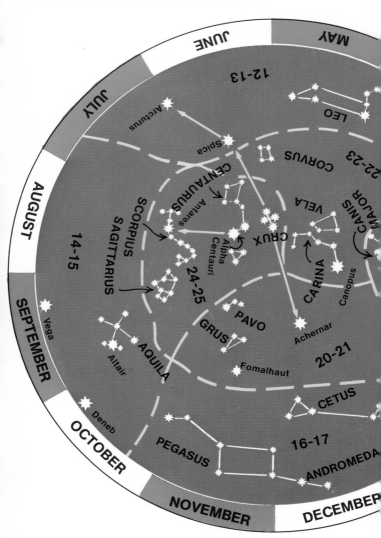

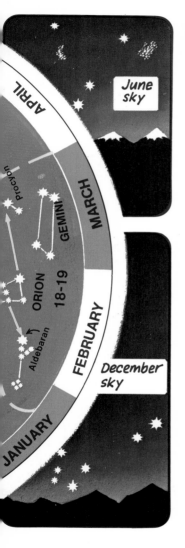

If you go to the southern hemisphere, use the map on the left to find your way around the skies. Turn the map until the present month is lowest. When you sit facing north you will be able to spot most of the stars shown in the center and lower part of the map.

When you look at Sagittarius, you are also looking toward the center of the galaxy, so you will see lots of stars in that area of the sky.

To the south you can see the southern cross, Crux. It can be used as a "signpost" to other constellations, as shown by the yellow arrows.

When you have found the prominent constellations shown here, turn to the page numbers marked on the map. The areas covered by each detailed map are separated by broken lines.

How to use the star charts on the following pages

Start by identifying the brightest stars and most prominent constellations. The size of the star symbols shows how bright each star is, not its actual size. You may have to tilt the maps to match them up with the sky. When you have recognized the obvious constellations, start looking for the fainter ones. It will probably take several nights before you know the sky well enough to spot them all.

Note that all the patterns will seem bigger in the sky than they appear on the maps. The two small views on the left show how the view of the sky changes at a particular spot through the year.

Draco to Cancer

1 Draco (dragon)
Long, straggling line of faint stars. Its "head" is a group of four stars near Vega; the "tail" loops around Ursa Minor. 3,000 years ago, Thuban was the pole star; today it is Polaris.

2 Canes Venatici (hunting dogs)
A constellation named in 1690. The "dogs" hunt the "bears," following them across the sky.

3 Boötes (herdsman)
A conspicuous kite-shape. Arcturus is the fourth brightest star in the sky: find it by using the curved handle of the Big Dipper as a pointer.

4 Coma Berenices (Berenice's hair)
A cloud of faint stars; binoculars will show about 30.

5 Virgo (virgin)
A constellation representing the goddess of justice. Five stars form an obvious "bowl." Spica is a hot, bright, white star. Its name means "ear of corn."

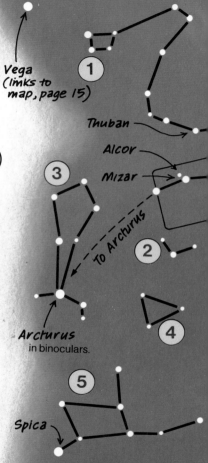

Vega (links to map, page 15)

Thuban

Alcor

Mizar

To Arcturus

Arcturus in binoculars.

Spica

12

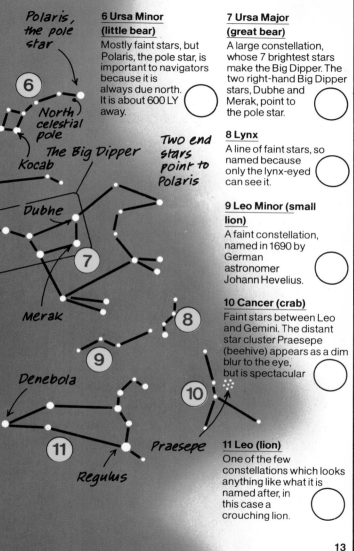

Polaris, the pole star

North celestial pole

Kocab

The Big Dipper

Dubhe

Merak

Denebola

Praesepe

Regulus

Two end stars point to Polaris

6 Ursa Minor (little bear)

Mostly faint stars, but Polaris, the pole star, is important to navigators because it is always due north. It is about 600 LY away.

7 Ursa Major (great bear)

A large constellation, whose 7 brightest stars make the Big Dipper. The two right-hand Big Dipper stars, Dubhe and Merak, point to the pole star.

8 Lynx

A line of faint stars, so named because only the lynx-eyed can see it.

9 Leo Minor (small lion)

A faint constellation, named in 1690 by German astronomer Johann Hevelius.

10 Cancer (crab)

Faint stars between Leo and Gemini. The distant star cluster Praesepe (beehive) appears as a dim blur to the eye, but is spectacular

11 Leo (lion)

One of the few constellations which looks anything like what it is named after, in this case a crouching lion.

13

Cygnus to Serpens

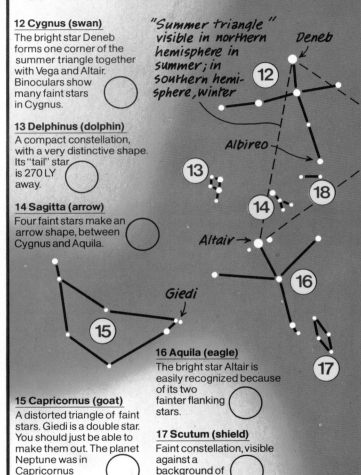

12 Cygnus (swan)
The bright star Deneb forms one corner of the summer triangle together with Vega and Altair. Binoculars show many faint stars in Cygnus.

13 Delphinus (dolphin)
A compact constellation, with a very distinctive shape. Its "tail" star is 270 LY away.

14 Sagitta (arrow)
Four faint stars make an arrow shape, between Cygnus and Aquila.

"Summer triangle" visible in northern hemisphere in summer; in southern hemisphere, winter

Deneb

Albireo

Altair →

Giedi

15 Capricornus (goat)
A distorted triangle of faint stars. Giedi is a double star. You should just be able to make them out. The planet Neptune was in Capricornus when discovered.

16 Aquila (eagle)
The bright star Altair is easily recognized because of its two fainter flanking stars.

17 Scutum (shield)
Faint constellation, visible against a background of the Milky Way.

14

18 Vulpecula (fox)

A very inconspicuous star group; originally called the fox and goose.

19 Lyra (lyre)

Small but easily spotted group. Vega is the fifth brightest star in the sky and 26 LY distant.

21 Hercules

A large constellation, but rather shapeless and difficult to recognize.

22 Corona Borealis (northern crown)

A semicircle of faint stars between Vega and Arcturus.

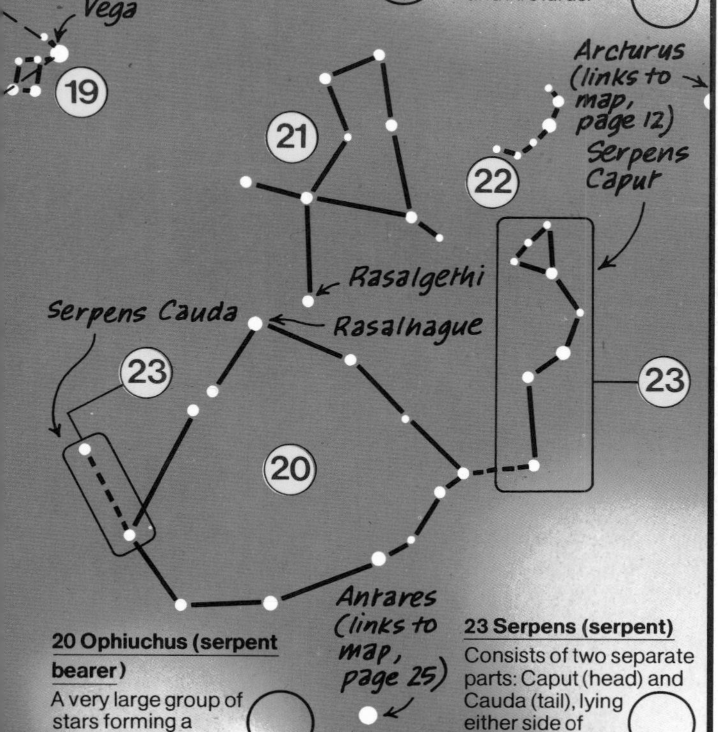

Vega

19

21

22

Arcturus (links to map, page 12)

Serpens Caput

Rasalgethi

Rasalhague

Serpens Cauda

23

23

20

20 Ophiuchus (serpent bearer)

A very large group of stars forming a distorted circle.

Antares (links to map, page 25)

23 Serpens (serpent)

Consists of two separate parts: Caput (head) and Cauda (tail), lying either side of Ophiuchus.

15

Camelopardalis to Aquarius

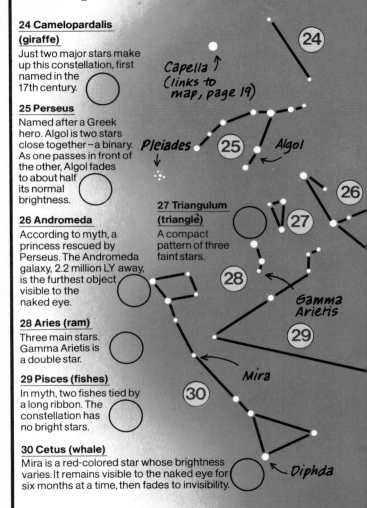

24 Camelopardalis (giraffe)

Just two major stars make up this constellation, first named in the 17th century.

25 Perseus

Named after a Greek hero. Algol is two stars close together – a binary. As one passes in front of the other, Algol fades to about half its normal brightness.

26 Andromeda

According to myth, a princess rescued by Perseus. The Andromeda galaxy, 2.2 million LY away, is the furthest object visible to the naked eye.

28 Aries (ram)

Three main stars. Gamma Arietis is a double star.

29 Pisces (fishes)

In myth, two fishes tied by a long ribbon. The constellation has no bright stars.

27 Triangulum (triangle)

A compact pattern of three faint stars.

30 Cetus (whale)

Mira is a red-colored star whose brightness varies. It remains visible to the naked eye for six months at a time, then fades to invisibility.

Capella ↑ (links to map, page 19)

Pleiades ↓

Algol ←

Gamma Arietis

Mira ←

Diphda ←

16

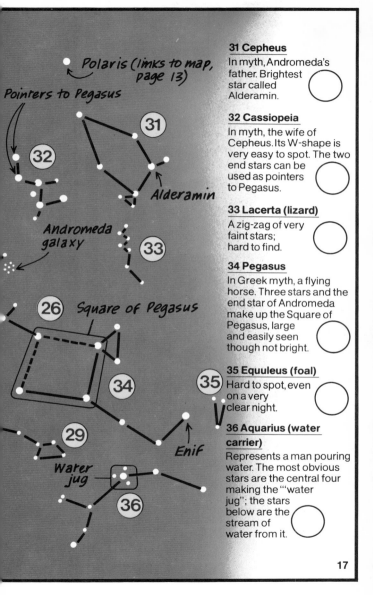

Polaris (links to map, page 13)

Pointers to Pegasus

31

32

Alderamin

Andromeda galaxy

33

26 Square of Pegasus

34

29

Enif

Water jug

36

31 Cepheus
In myth, Andromeda's father. Brightest star called Alderamin.

32 Cassiopeia
In myth, the wife of Cepheus. Its W-shape is very easy to spot. The two end stars can be used as pointers to Pegasus.

33 Lacerta (lizard)
A zig-zag of very faint stars; hard to find.

34 Pegasus
In Greek myth, a flying horse. Three stars and the end star of Andromeda make up the Square of Pegasus, large and easily seen though not bright.

35 Equuleus (foal)
Hard to spot, even on a very clear night.

36 Aquarius (water carrier)
Represents a man pouring water. The most obvious stars are the central four making the "'water jug"; the stars below are the stream of water from it.

17

Gemini to Lepus

37 Gemini (twins)

Castor is actually six stars very close together, but binoculars cannot separate them. The faint planets Uranus and Pluto were in Gemini when discovered.

38 Canis Minor
(small dog)

In myth, the smaller of the two dogs of Orion the hunter. Procyon is the eighth brightest star in the sky, and at 11 LY, among the closest to Earth.

39 Monoceros
(unicorn)

Inconspicuous and recently named (in the 17th century), but worth "sweeping" with your binoculars for star clusters and nebulae.

40 Canis Major
(large dog)

A compact group of bright stars. Sirius (the Dog Star) is the brightest star in the sky, and only 8 LY away from Earth. Read more about it on page 29.

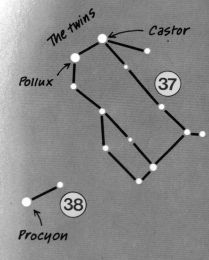

The twins
Castor
Pollux
37

38
Procyon

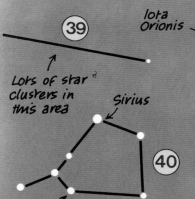

39
Iota Orionis
Lots of star clusters in this area
Sirius
40
Adara

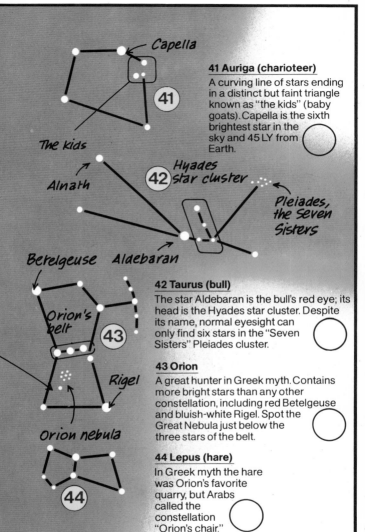

41 Auriga (charioteer)

A curving line of stars ending in a distinct but faint triangle known as "the kids" (baby goats). Capella is the sixth brightest star in the sky and 45 LY from Earth.

42 Taurus (bull)

The star Aldebaran is the bull's red eye; its head is the Hyades star cluster. Despite its name, normal eyesight can only find six stars in the "Seven Sisters" Pleiades cluster.

43 Orion

A great hunter in Greek myth. Contains more bright stars than any other constellation, including red Betelgeuse and bluish-white Rigel. Spot the Great Nebula just below the three stars of the belt.

44 Lepus (hare)

In Greek myth the hare was Orion's favorite quarry, but Arabs called the constellation "Orion's chair."

Columba to Microscopium

45 Columba (dove)
A distinct group lying near Canopus, named in 1679.

46 Horologium (pendulum clock)
Only one star is easily visible in this constellation.

48 Reticulum (net)
A distinct little group of faint stars between Canopus and Achernar.

49 Mensa (Table Mountain)
The faintest constellation in the sky. Look for it on a very clear night.

50 Hydrus (small water snake)
A large triangle lying between the misty patches of the Magellanic Clouds.

51 Octans (octant)
Always due south, but has no bright star to match Polaris in the northern hemisphere.

52 Pavo (peacock)
Conspicuous group of faint stars, named in 1603. Kappa Pavonis is a variable star, changing from dim to bright and back again every 9.1 days.

Lower part of Orion (links to map, page 19)

47 Caelum (chisel)
Consists of a few very faint stars.

Rigel

Canopus (links to map, page 23)

Large Magellanic Cloud

South celestial pole

Apus

Small Magellanic Cloud

Kappa Pavonis

20

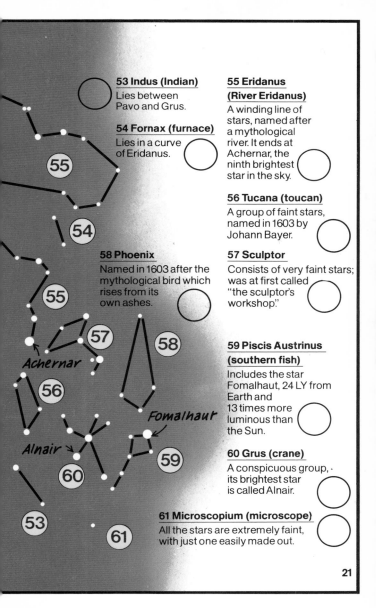

53 Indus (Indian)
Lies between Pavo and Grus.

54 Fornax (furnace)
Lies in a curve of Eridanus.

58 Phoenix
Named in 1603 after the mythological bird which rises from its own ashes.

55 Eridanus (River Eridanus)
A winding line of stars, named after a mythological river. It ends at Achernar, the ninth brightest star in the sky.

56 Tucana (toucan)
A group of faint stars, named in 1603 by Johann Bayer.

57 Sculptor
Consists of very faint stars; was at first called "the sculptor's workshop".

59 Piscis Austrinus (southern fish)
Includes the star Fomalhaut, 24 LY from Earth and 13 times more luminous than the Sun.

60 Grus (crane)
A conspicuous group, its brightest star is called Alnair.

61 Microscopium (microscope)
All the stars are extremely faint, with just one easily made out.

Achernar

Fomalhaut

Alnair

Corvus to Dorado

62 Corvus (crow)
A distinct foursome of stars in a rather barren area of the sky.

63 Crater (cup)
Another group of four stars, like a fainter Corvus.

64 Antlia (air pump)
A triangle of faint stars, named in 1763.

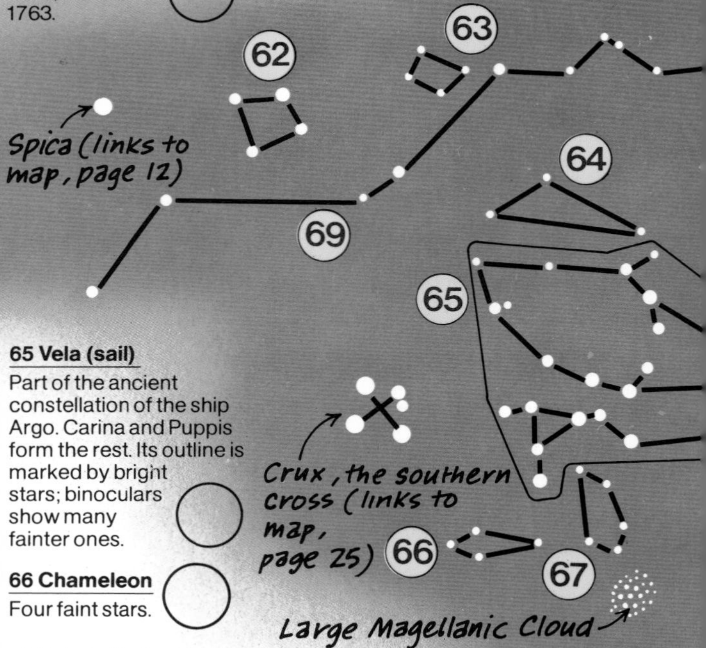

Spica (links to map, page 12)

65 Vela (sail)
Part of the ancient constellation of the ship Argo. Carina and Puppis form the rest. Its outline is marked by bright stars; binoculars show many fainter ones.

66 Chameleon
Four faint stars.

67 Volans (flying fish)
Distinct group of faint stars, partly enclosed by Carina.

Crux, the southern cross (links to map, page 25)

Large Magellanic Cloud

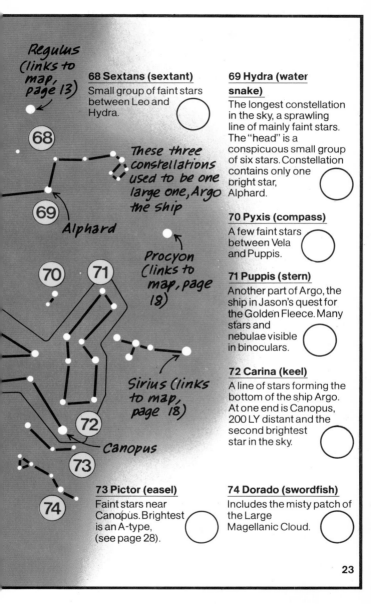

Regulus (links to map, page 13)

68 Sextans (sextant)
Small group of faint stars between Leo and Hydra.

These three constellations used to be one large one, Argo the ship

Alphard

Procyon (links to map, page 18)

Sirius (links to map, page 18)

Canopus

69 Hydra (water snake)
The longest constellation in the sky, a sprawling line of mainly faint stars. The "head" is a conspicuous small group of six stars. Constellation contains only one bright star, Alphard.

70 Pyxis (compass)
A few faint stars between Vela and Puppis.

71 Puppis (stern)
Another part of Argo, the ship in Jason's quest for the Golden Fleece. Many stars and nebulae visible in binoculars.

72 Carina (keel)
A line of stars forming the bottom of the ship Argo. At one end is Canopus, 200 LY distant and the second brightest star in the sky.

73 Pictor (easel)
Faint stars near Canopus. Brightest is an A-type, (see page 28).

74 Dorado (swordfish)
Includes the misty patch of the Large Magellanic Cloud.

23

Sagittarius to Crux

75 Sagittarius (archer)

A distinctive "teapot" shape of bright stars. The misty nebula M8 is a region where stars are forming. Many other nebulae are visible with binoculars.

76 Corona Australis
(southern crown)

Faint stars in a curving group.

77 Telescopium (telescope)

A group of faint stars near the "sting" of Scorpius.

78 Ara (altar)

Lies between Alpha Centauri and the "sting" of Scorpius.

79 Circinus (compasses)

Consists of three faint stars near Alpha Centauri; named in the 18th century.

80 Triangulum Australe (southern triangle)

An easily spotted triangle of bright stars, named in 1603. The brightest star in the group is 100 LY away.

81 Apus (bird of paradise)

An inconspicuous group of faint stars, named in 1603.

82 Musca (fly)

A constellation of faint stars next to the southern cross.

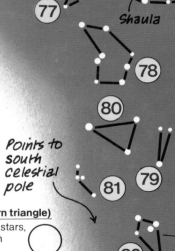

Nunki

M8 nebula

Shaula

Points to south celestial pole

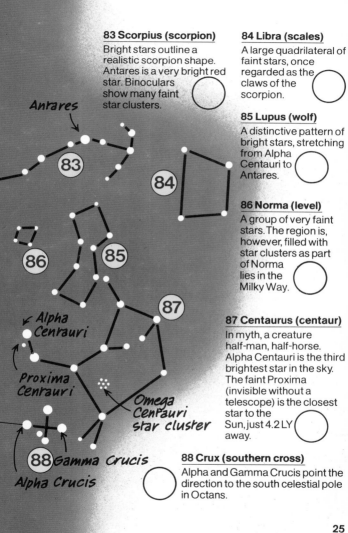

83 Scorpius (scorpion)

Bright stars outline a realistic scorpion shape. Antares is a very bright red star. Binoculars show many faint star clusters.

84 Libra (scales)

A large quadrilateral of faint stars, once regarded as the claws of the scorpion.

85 Lupus (wolf)

A distinctive pattern of bright stars, stretching from Alpha Centauri to Antares.

86 Norma (level)

A group of very faint stars. The region is, however, filled with star clusters as part of Norma lies in the Milky Way.

87 Centaurus (centaur)

In myth, a creature half-man, half-horse. Alpha Centauri is the third brightest star in the sky. The faint Proxima (invisible without a telescope) is the closest star to the Sun, just 4.2 LY away.

88 Crux (southern cross)

Alpha and Gamma Crucis point the direction to the south celestial pole in Octans.

Antares

Alpha Centauri

Proxima Centauri

Omega Centauri star cluster

88 Gamma Crucis
Alpha Crucis

Stellar birthplaces

Stars are formed from the very tenuous hydrogen and helium gas and dust which fills space. Denser clouds of gas are called nebulae. Within them, gravitation condenses and heats up the gas until stars are formed – huge balls of hot gas, a million or more miles across. At the center of a star like the Sun the temperature is about 15 million° C; it burns by nuclear reaction like a slow-motion H-bomb.

These pages show some nebulae and star clusters. There are millions more, but the ones shown here are visible as faint, fuzzy patches to the naked eye.

Some star clusters stay together, but many break up. The stars may end up single, like the Sun, or more often in a pair or a trio.

▲ Lagoon nebula, M8 (page 24)

Large clouds made of hydrogen and helium gas and dust. The star that makes the gas glow is so deeply embedded in dust that it cannot be seen. Nebula is 4,580 LY away.

◀ Praesepe (page 13)

An easily spotted cluster, 18 LY across and about 525 LY from Earth. Most of the stars (about 100) are thought to be about 400 million years old. No glowing nebula of hydrogen gas is visible, even through powerful telescopes.

▲ Pleiades (page 19)

A cluster of about 200 stars formed about 60 million years ago. Often called the "Seven Sisters," though only six are visible to the naked eye.

◄ Omega Centauri (page 25)

Only visible from the southern hemisphere. The cluster consists of a million stars.

Orion Nebula (pages 1 and 19)

Visible just below Orion's belt. The nebula is about 20 LY across and 1,500 LY from Earth.

Types of stars

The Sun is a typical star, but not all stars are like the Sun – they vary enormously in size, color and temperature.

Newly formed stars like the ones on the previous page cover a wide range, from extremely bright and hot, bluish-white stars to dim, cooler ones. The Sun is a "middle-aged" star, 5,000 million years old.

Over 30 percent of known stars are binary and many stars vary in brightness, unlike the Sun, whose light remains steady.

The color of stars varies from an intense blue-white through yellow and orange to a dim red. The shade indicates temperature – the cooler the star, the redder it appears. This chart plots typical colors and temperatures, together with stars of each kind.

Color				
Blue-White	White	Yellow	Orange	Red
Surface Temperature in degrees Centigrade				
25,000	10,000	6,000	4,000	3,000
Typical star				
Spica	Sirius	Sun	Arcturus	Betelgeuse

Most stars are classified into the seven groups shown below. Each group is divided into a further ten subdivisions – the Sun is spectral type G2. The types are arranged mainly by an analysis of a star's light, temperature and chemical compounds.

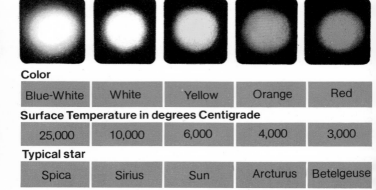

	O	B	A	F	G	K	M
30	← Iota Orionis						
25							
20		Rigel					
15			Vega	Canopus	Sun Aldebaran		
10							Antares
5,000°C							
Type	**O**	**B**	**A**	**F**	**G**	**K**	**M**

◀ **Variable star Algol (page 16)**

Every three days, this star dims for ten hours but not because its light output changes. It is in fact a double star, and one of the pair periodically blocks off the light of the other, reducing the apparent brightness. The bright star is type B8, the fainter star type K.

Check off if you can spot Algol's brightness change

Sirius (page 18) ▶

Sirius, 8.6 LY from Earth and spectral type A1, has a faint companion which is a white dwarf, a collapsed star core. A spoonful of matter from it would weigh 20 tons. You cannot see the dwarf, but should have no problem finding Sirius. Sirius is known as the "Dog Star" as it is in Canis Major. It is the brightest star in the sky.

◀ **Mizar and Alcor (page 12)**

Mizar, the second star in the "handle" of the Big Dipper, has a companion star, Alcor, which can be spotted with the naked eye. Mizar itself is a double star, but the pair can only be seen with a telescope. Alcor is arrowed in the picture. Mizar and Alcor are about 80 LY away.

Mizar *Alcor*

Dying stars

A star does not shine forever. The hydrogen fuel at its center is eventually exhausted as the "hydrogen bomb" nuclear reaction turns it all to helium (comparable to the ash remaining after a coal or wood fire). At this point the star swells up to a hundred times its previous size. Huge stars like this are red – and some of the brightest stars in the sky are "red giants".

It then begins to burn helium as a nuclear fuel. Eventually, as its central nuclear fuel approaches complete exhaustion, a red giant blows up, its outer layers expanding into space. The central core collapses in upon itself, cools down and fades away. Dead stars like this are very small and very dense; their collapsed matter makes lead seem as light as a feather.

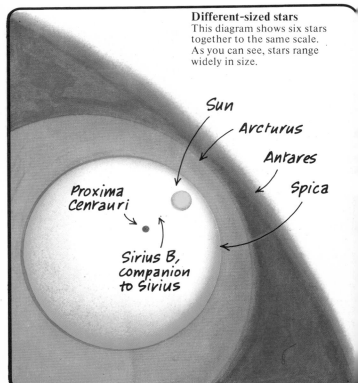

Different-sized stars
This diagram shows six stars together to the same scale. As you can see, stars range widely in size.

Sun

Arcturus

Antares

Spica

Proxima Cenrauri

Sirius B, companion to Sirius

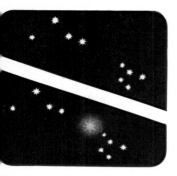

◀ Mira (page 16)

The central star in the constellation of Cetus is a red giant spectral type M whose brightness varies. Over months its brightness can increase about a thousand times. It is known as a "long-period" variable star because its brightness change takes a little under a year.

Antares (page 25) ▶

Another old red giant, about 250 million miles in diameter. It has a small companion star, type B4, that can only be seen with a telescope. As you can see from the picture, it is a greenish color, but this is an optical effect caused by the contrast with bright red Antares. A cool star, Antares' surface temperature is only 3,200°C.

◀ Betelgeuse (page 19)

A red giant, easily spotted at the "shoulder" of Orion. This picture is based on a computer analysis of temperature variations on the surface of Betelgeuse. The star is a variable – try comparing it in the sky to Aldebaran to spot slow changes in brightness. Its size varies with its brightness. At its maximum, it is nearly 350 million miles across, about 400 times larger than the Sun. It is an M2 spectral type.

31

Galaxies

The stars you see in the sky (including the Sun) are part of a huge spiral group of 100,000 million stars, called the galaxy.

Although you can see only a fraction of these stars, the light from the distant ones blends together to form a misty glowing band stretching right across the sky. Its name is the Milky Way and it passes through the constellation of Cassiopeia, Cygnus, Aquila, Sagittarius, Scorpius, Centaurus, Crux, Vela, Puppis, Monoceros and Perseus.

Far out beyond our own Milky Way galaxy, there are billions more. Most galaxies are so distant that a telescope is needed to see them but three are visible to the naked eye.

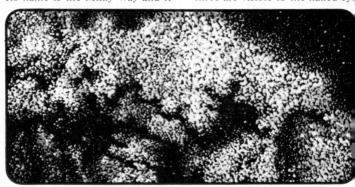

▲ Milky Way

This is a section of the Milky Way as seen through a telescope. Using binoculars you can see many faint nebulae and star clusters within it.

◄ Coal Sack

Dust particles in space absorb starlight, producing dark "holes" in the Milky Way. Look near Crux to find the "Coal Sack" shown here.

◀ Magellanic Clouds
(pages 20, 21, 22, 23)

These bright misty patches, only visible from the southern hemisphere, were first spotted by the explorer Ferdinand Magellan in 1521. They are smaller than our own galaxy and are the closest galaxies to the Milky Way. Neither has a distinctive shape, which is unusual – larger galaxies are spiral or oval in shape.

Large Magellanic Cloud (top picture) →◯

Small Magellanic → ◯ Cloud

◀ Andromeda Galaxy
(page 17)

To the naked eye it looks like a faint blur. It is 2.2 million LY away and is the most distant object you can see without binoculars or telescope. Its spiral shape is similar to that of the Milky Way. Galaxies group together in clusters. Andromeda is a member of the Local Group along with the Milky Way, the Magellanic Clouds and 30 other galaxies. ◯

Empire of the Sun

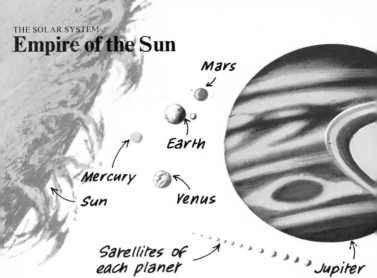

Mars

Earth

Mercury

Sun

Venus

Satellites of each planet

Jupiter

On a clear night, you will probably see one or more bright points of light which are not marked on the star maps. These are the planets.

The planets gradually change their positions against the background stars, so they cannot be marked on the constellation maps.

After each planet's description on the following pages there is a table showing which constellation it is moving through during the next few years. Apart from Uranus, Neptune and Pluto, the planets appear very bright, and once you have located the constellations you

Planetary fact-finder		
Planet	**Diameter**	**Average distance from Sun**
Mercury	3,032 mi	36 million mi
Venus	7,500 mi	67 million mi
Earth	7,926 mi	93 million mi
Mars	4,217 mi	141 million mi
Jupiter	88,700 mi	482 million mi
Saturn	74,500 mi	885 million mi
Uranus	32,200 mi	1,780 million mi
Neptune	30,750 mi	2,800 million mi
Pluto	2,500 mi	3,660 million mi

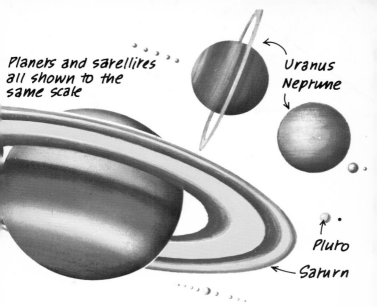

Planets and satellites all shown to the same scale

Uranus
Neptune
Pluto
Saturn

cannot miss them.

Like the Earth, the other planets circle the Sun. They have no light of their own, and only shine because they reflect sunlight.

All the planets revolve around the Sun in the same direction, and their orbits lie nearly in the same plane like the tracks on an LP record. As a result, all the planets seem to move through the same group of constellations, the zodiac. The zodiac consists of Pisces, Aries, Taurus, Gemini, Cancer, Leo, Virgo, Libra, Scorpius, Sagittarius, Capricornus and Aquarius.

Time to orbit Sun (year)	Time to rotate (day)	Number of satellites
88 days	59 days	None
225 days	243 days	None
365.3 days	23 hrs 56 mins	1 (the Moon)
687 days	24 hrs 37 mins	2
11.9 years	9 hrs 50 mins	13 or 14
29.5 years	10 hrs 14 mins	9 or 10
84 years	24 hrs (?)	5
165 years	15 hrs (?)	2
248 years	6 days 10 hrs	1

Mercury

Mercury, the closest planet to the Sun, is only 50 percent larger than the Moon. It has no atmosphere, and the Sun bakes its surface during its day, up to 350°C, while it becomes bitterly cold, −170°C, at night. The planet is heavily cratered like the Moon, and it has long ridges where it has shrunk slightly, like an old apple. This picture is based on photographs taken by the Mariner 10 spacecraft in March 1974.

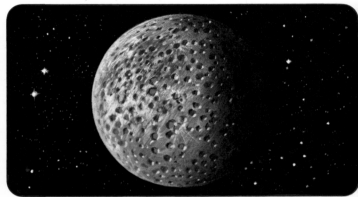

Mercury is not very easy to spot. Being always close to the Sun, it is only seen low on the horizon in the twilight glow.

Mercury – when and where to look			
Year	**East before sunrise**	**Year**	**West after sunset**
1979	Late April	1979	Early March, late Oct.
1980	Early April	1980	Early October
1981	Mid March, late Oct.	1981	Late September
1982	Early March, mid Oct.	1982	Early May, early Sept.
1983	Early October	1983	Mid-late April
1984	Mid May, mid Sept.	1984	Early April, mid Nov.

Venus

Venus is almost as large as Earth, but its atmosphere is a hundred times thicker and composed of choking carbon dioxide. This thick blanket makes Venus's surface very hot – 480°C – hot enough to melt lead. Venus is veiled by continuous clouds, probably made of sulfuric acid droplets. Radar experiments and space probes have shown that its surface has craters, mountains and valleys.

Venus is the brightest object in the sky after the Sun and Moon. It appears either as the "Evening Star" after sunset, or the "Morning Star" before sunrise.

Venus – when and where to look

Year	East before sunrise	Year	West after sunset
1979	January – August	1979	September – December
1980	June – December	1980	January-May
1981	January – March	1981	April – December
1982	February – October	1982	January, then December
1983	September – December	1983	January – August
1984	January – June	1984	July – December

Mars

Mars is a rocky planet half the size of Earth. It has a very thin atmosphere, of unbreathable carbon dioxide, and has no liquid water on the surface. Its entire surface, apart from icy polar caps, is a dry red desert. Despite its small size, Mars has canyons and volcanos larger than any on Earth: one volcano, Olympus Mons, is 15 miles high and 310 miles across. Mars has two small moons, Phobos and Deimos.

This picture is based on photographs taken by American spaceprobes that have landed on the planet and sampled the soil and air. Mars appears in the sky as a bright, red "star."

Mars – where and when to look			
Year	**Months visible**	**Constellation**	**Page**
1979	October November–December	Cancer Leo	13 13
1980	January–June	Leo	13
1981	October–November December	Leo Virgo	13 12
1982	January–August September	Virgo Libra	12 25
1983	November–December	Virgo	12
1984	January–July August–September	Libra Scorpius	25 24–25

Jupiter

Jupiter is the largest planet, but despite its size, Jupiter spins faster than the Earth. Its day lasts less than 10 hours.

The stripes visible in a telescope are cloud layers. Its most noticeable feature is the Great Red Spot shown below left. This is a giant eddy in Jupiter's atmosphere; its center is fairly calm. It was first spotted in 1665 and was thought to be a volcano.

Jupiter appears as a bright, yellowish white point of light, brighter than any of the stars. On the left is the Great Red Spot compared in size to the Earth.

Jupiter – where and when to look

Year	Months visible	Constellation	Page
1979	January–July	Cancer	13
	August–December	Leo	13
1980	January–September	Leo	13
	October–December	Virgo	12
1981	January–December	Virgo	12
1982	January–December	Libra	25
1983	January–December	Scorpius	24-25
1984	January–December	Sagittarius	24

Saturn

Saturn is the second largest planet and consists almost entirely of substances familiar on Earth as gases: methane, hydrogen and ammonia. On Saturn these are compressed by gravity to be a liquid. If there is any rock, it forms only a tiny core at the planet's center.

Saturn is famous for its rings. These consist of billions of rock and ice particles, all orbiting Saturn like miniature moons.

The telescope view above shows Saturn and its rings. It appears to the naked eye as a bright yellow "star." Its rings are only visible through a telescope.

Saturn – where and when to look			
Year	**Months visible**	**Constellation**	**Page**
1979	January–October	Leo	13
	November–December	Virgo	12
1980	January–December	Virgo	12
1981	January–December	Virgo	12
1982	January–December	Virgo	12
1983	January–October	Virgo	12
	November–December	Libra	25
1984	January–December	Libra	25

Uranus, Neptune, Pluto

None of the three outer planets is visible to the naked eye. Little is known of them as details are difficult to see even with powerful telescopes.

▼ Uranus

Discovered in 1781 by William Herschel. The picture below shows its faint ring system which was discovered in 1977.

Neptune ▶

Bluish in color, the planet is similar to Uranus in size. It has two satellites, Nereid and Triton. Triton may be the largest satellite in the Solar System, nearly 3,800 miles in diameter.

Pluto ▶

These two pictures were taken in 1930. One of the "stars" (arrowed) moved, showing that it was a planet. Only powerful telescopes can show Pluto and no details of its surface can be seen. Pluto's orbit passes inside that of Neptune, and until 1999, Neptune will be the outermost planet.

Comets

Comets are balls of ice, dust and rock, drifting in huge elongated orbits that extend far into space, even beyond the orbit of Pluto. As a comet nears the Sun, the heat turns the ice into a mini-atmosphere, that streams out into a tail extending for millions of miles. Sometimes a comet is large and bright enough to be seen without binoculars, though most are only visible through telescopes.

Tail of gas and dust – can be millions of miles long

Nucleus – made of rocks and ice

Coma – made of gases

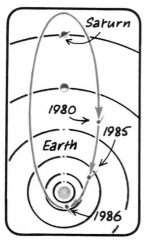

Saturn

1980

1985

Earth

1986

◄ Halley's Comet

Most comets appear unexpectedly and will not return for thousands of years.
A regular visitor is Halley's Comet which returns to the Sun every 76 years.
The diagram shows its orbit.
It will next be visible in February 1986, for perhaps two weeks.

This picture shows Ikeya-Seki, a comet which appeared in 1965 →

Asteroids

Between the orbits of Mars and Jupiter there are thousands of rocks, called asteroids, orbiting the Sun. They range in size from 623 miles downward. Even the largest are too faint to be seen with the unaided eye. The early Solar System contained only asteroidlike rocks. Most of them accumulated to make the planets, leaving the remainder orbiting in a belt between Mars and Jupiter.

Asteroid Belt

Pallas

Ceres

Moon

Vesta

Over 2,000 asteroids have precisely known orbits, and have been given individual names. It is estimated that 50,000 asteroids are visible with a large telescope. The picture on the left shows the three largest compared in size with the Moon.

The largest asteroids	
Ceres	623 mi dia
Pallas	378
Vesta	334
Hygeia	280
Euphrosyne	230

Pioneer 10, the first spaceprobe to cross the Asteroid Belt, on its way to Jupiter

The Moon

The Moon is Earth's only natural satellite. It is an airless, dry and small world, just 2,160 miles in diameter. This is large for a satellite, however, and many astronomers regard Earth as a double-planet system, unique in the Solar System.

The Moon's surface is covered with round craters, up to 155 miles across, that were blasted out by asteroids and meteorites millions of years ago. The Moon has high mountains, some almost 33,000 feet high.

Here are nineteen lunar features which are easy to spot. You will need binoculars to see ones marked ★

1 **Mare Tranquillitatis** Sea of Tranquility

2 **Mare Serenitatis** Sea of Serenity

3 **Mare Humorum** Sea of Humors

4 **Mare Vaporum** Sea of Vapors

5 **Mare Frigoris** Sea of Cold

6 **Mare Imbrium** Sea of Showers

7 **Mare Nubium** Sea of Clouds

8 **Mare Crisium** Sea of Crises

9 **Mare Nectaris** Sea of Nectar

The farside of the Moon ⟶

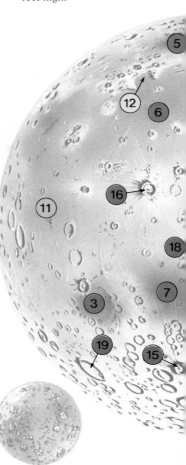

The dark patches spread over large areas are plains of solidified lava. Early astronomers thought they were seas and oceans and they still bear the Latin names *mare* (sea), *oceanus* (ocean) and *sinus* (bay).

The Moon always keeps the same side to Earth. It turns on its own axis in exactly the same time as it takes to orbit the Earth, 27⅓ days. No one knew what the other side looked like until space probes took pictures in the 1950s and 60s.

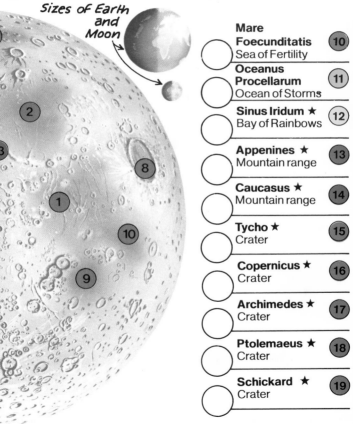

Sizes of Earth and Moon

Mare Foecunditatis (10)
Sea of Fertility

Oceanus Procellarum (11)
Ocean of Storms

Sinus Iridum ★ (12)
Bay of Rainbows

Appenines ★ (13)
Mountain range

Caucasus ★ (14)
Mountain range

Tycho ★ (15)
Crater

Copernicus ★ (16)
Crater

Archimedes ★ (17)
Crater

Ptolemaeus ★ (18)
Crater

Schickard ★ (19)
Crater

Phases of the Moon

The Full Moon is the brightest object in the sky after the Sun, but it emits no light of its own. Like the planets, it merely reflects sunlight. As the Moon orbits Earth, different amounts of the sunlit half are visible so the lit shape, (called the phase) changes. The phases repeat every 29½ days. Although it looks so bright, the Moon only reflects about seven percent of the sunlight falling on it – its surface is quite dark.

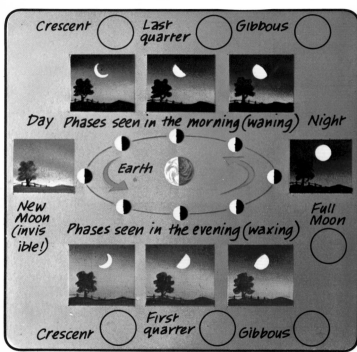

Crescent · Last quarter · Gibbous

Day · Phases seen in the morning (waning) · Night

Earth

New Moon (invisible!)

Phases seen in the evening (waxing)

Full Moon

Crescent · First quarter · Gibbous

Lunar facts and figures

Diameter	2,160 miles
Average distance from Earth	238,600 miles
Time taken to orbit Earth	27.3 days
Time to rotate once (lunar day)	27.3 days
Surface temperature	Day 105°C Night −155°C

Eclipses

he New Moon sometimes passes
 front of the Sun, cutting off its
ght. This is called an eclipse of the
.n. When Full, the Moon some-
mes passes into the shadow of the
arth. There is little light to reflect,

so the Moon dims to a very faint
copper color. This is called an
eclipse of the Moon.

The two diagrams on this page
show how eclipses occur, but they
are not to scale.

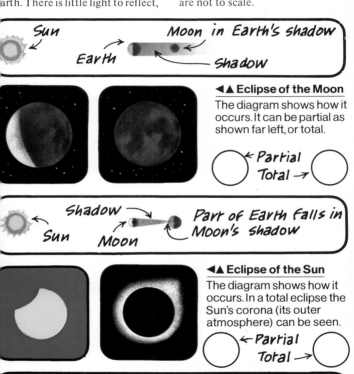

◀▲ Eclipse of the Moon
The diagram shows how it
occurs. It can be partial as
shown far left, or total.

← Partial
Total →

◀▲ Eclipse of the Sun
The diagram shows how it
occurs. In a total eclipse the
Sun's corona (its outer
atmosphere) can be seen.

← Partial
Total →

Danger – do not look at the Sun!
Many people think smoked glass protects your eyes if you want
to look at the Sun. It doesn't. Make a viewer like the one on
page 55 – it works well and provides you with a magnified image
to look at in comfort.

Meteors

In the space between planets there is assorted debris ranging from rocks
miles across to tiny grains of dust. These particles, which can move at up
to 20 miles per second, are called meteoroids.

Meteors

When a meteoroid collides with
the Earth's atmosphere, friction
heats it white-hot and it shines
briefly as a meteor or "shooting
star" before it burns up.
A really bright one is called a
fireball; it usually leaves a briefly
glowing trail.

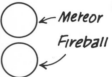

← Meteor

Fireball ←

Meteor showers

Small meteoroids are the remains
of broken-up comets. As the Earth
crosses various comet orbits there
are spectacular showers of meteors.

The effect of perspective makes
the meteors seem to spread out
from a point in the sky, just as a
road seems to spread apart from
a distant point. Meteor showers are
named after the constellation in
which the spreading-out point, or
radiant lies.

Most meteors become visible
some distance away from the
radiant, so when spotting a meteor
shower do not look directly at the
radiant. When you see a meteor
trace its path backwards in the sky.
If this line goes through the constellation containing the radiant you
have seen a shower meteor.

Meteorites

If a meteor crashes into the
ground it is called a meteorite.
To see one land is very, very rare
so check off the circles if you
see meteorites in a museum —
many have collections.

▲ Stony meteorites

There are two main types of
meteorites — stony and iron.
The technical name for
a stony one is an aerolite.
Often covered with a
smooth black crust.

Iron meteorite ▶

Known as a siderite. Examination
under a microscope reveals a
peculiar crisscross pattern.
A million-ton siderite
blasted a 3,600 feet-
wide hole in Arizona
50,000 years ago.

Important meteor showers

Dates visible	Name	Radiant	
January 1-6	Quadrantids	Boötes (page 12)	
April 19-24	Lyrids	Lyra (page 15)	
May 1-8	Eta Aquarids	Aquarius (page 17)	
July 25 – August 18	Perseids	Perseus (page 16)	
October 16-21	Orionids	Orion (page 19)	
October 20 – November 30	Taurids	Taurus (page 19)	
December 7-15	Geminids	Gemini (page 18)	

Other sky sights

In addition to the stars and planets there are other things to spot, nearer the Earth. Note that aurorae cannot be seen from tropical latitudes; zodiacal light cannot be seen from polar latitudes.

◄ Aurora Borealis/Australis

Known as Borealis in northern hemisphere, Australis in south. Shimmering curtain effect is caused by radiation from Sun striking particles in Earth's upper atmosphere.

Zodiacal light ►

Faint cone of light caused by sunlight reflecting off dust particles in space. Best seen from tropics. Glow passes through the constellations of the zodiac.

◄ Nighttime vapor trails

The upper atmosphere remains in sunlight for a short period after sunset on the ground below. Aircraft vapor trails glow faintly as the ice crystals of which they are composed reflect the Sun.

Halo round the Moon ►

Ice particles in the upper atmosphere cause this effect. Usually one ring, but sometimes two, can be seen. Normally silvery white in color. Very occasionally, haloes can look like very pale rainbows.

Artificial satellites

Since the Space Age began in 1957 with the launch of Sputnik I, thousands of artifical satellites have been launched to orbit Earth. The largest satellites (and discarded rockets in orbit) can be seen with the unaided eye. They look like slow but steadily moving points of white light. Beware of mistaking aircraft for satellites. Aircraft generally have colored identification lights, and their engines are usually audible on a still night, while satellites are silent.

Artificial satellite ▶
Some newspapers carry details of the time that satellites are due, though there are so many that an hour's watch will usually enable you to spot one. As the satellite passes into the Earth's shadow it will be eclipsed, fading out of sight.

Artificial satellite → Eclipse ←

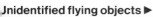

◀ Flashing satellite
This is not a light fitted on board, but the result of the satellite spinning in orbit. Sunlight reflects off different parts causing the flashing effect as it slowly moves across the sky.

Unidentified flying objects ▶
UFOs are unexplained moving lights – some people think they are spaceships from other worlds. About 40 sightings are reported across the world every day, but most are hoaxes or due to confusion with ordinary objects.

Taking photographs

You can take photographs of the sky if you have a camera with a "Brief Time" (B) setting. When set to this, the shutter stays open as long as you keep your finger pressed on the shutter release button.

If the camera has aperture and focus controls, use the smallest f number (2.8 is common) and focus on infinity (∞ is the symbol used). Color slide film is best. If you use negative or black and white film, make sure that whoever develops it knows they are star pictures. Otherwise they will think there is nothing on the film and give you no prints.

Wide aperture

B setting

Cable release

▲ Setting up the camera

It must be fixed firmly throughout the exposure. Prop the camera on a wall if you can, or better still, mount it on a tripod and use a cable release to press the button.

Star trails ▶

Point the camera at Polaris (or Octans in southern skies). Keep the shutter open several minutes (or hours). You should get a result like the one shown here as the polar stars rotate in the sky.

Constellation legends

Most of the prominent constellations have names taken from ancient Greek myths and legends. Many of these are based on the stories of still older civilizations.

On this page you can see just one group of constellations and their story. As you can see, the Greeks certainly used their imaginations when they related star patterns to human and animal shapes! There are many others – try finding out about them from an encyclopedia.

The Greeks saw the stars shown here, visible on January and February nights, as a giant display dominated by the great hunter Orion. He is facing a charging bull, Taurus, with a raised club and lion-skin shield. Behind him are his two faithful hunting dogs, Canis Major and Canis Minor. Unnoticed at his feet is his favourite quarry. Lepus the hare.

Taurus the bull

Canis Minor

Orion the hunter

Canis Major

Lepus the hare

Constellation quiz

Name each constellation and fill in the missing star in each pattern. Constellation names are shown at the bottom of the page.

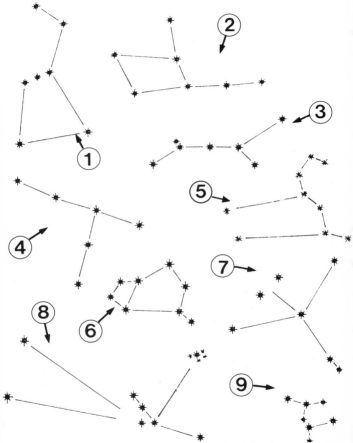

Answers: page numbers are given so you can check for the star names.

1 Orion (p19)	4 Cygnus (p14)	7 Aquila (p14)
2 Virgo (p12)	5 Leo (p13)	8 Taurus (p19)
3 Ursa Major (p13)	6 Sagittarius (p24)	9 Cassiopeia (p17)

54

Make a Sun-spotter

This is the only way to look at the Sun safely. Looking at it directly will blind you; filters or smoked glass are not safe either.

The Sun-spotter takes about 30 minutes to make and its results are excellent. Prop the spotter on a chair to get a good steady image.

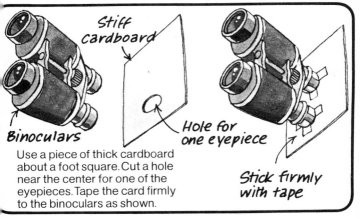

Stiff cardboard

Binoculars

Hole for one eyepiece

Stick firmly with tape

Use a piece of thick cardboard about a foot square. Cut a hole near the center for one of the eyepieces. Tape the card firmly to the binoculars as shown.

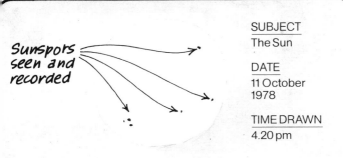

Sunspots seen and recorded

SUBJECT
The Sun

DATE
11 October
1978

TIME DRAWN
4.20 pm

Prop the binoculars on a chair so that they point to the Sun. Project the sun's image onto another piece of white cardboard placed near the eyepiece of the binoculars.

Focus them until the image is sharp. **Do not look through the binoculars.** Try tracking sunspots over several days – they move slowly as the Sun rotates.

Interplanetary puzzles

Name the planets, in order, outwards from the Sun.

1 _____

2 _____

3 _____

4 _____

5 _____

6 _____

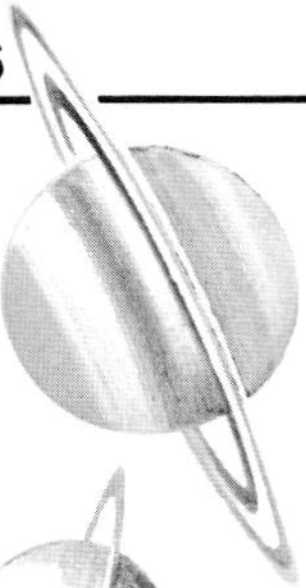

7 _____

8 _____

9 _____

Which planets have the following number of satellites?

Number of Satellites	Planet(s)
13 or 14	
5	
1	
2	
9 or 10	
None	

How far are these planets, on average, from the Sun?

	Million miles
Mercury	
Earth	
Jupiter	
Uranus	
Pluto	

Answers: You will find all the answers by referring back to pages 34 and 35.

Pronunciation guide

The names of many of the stars and constellations are hard to pronounce. In this list of the more difficult ones, emphasise the syllable printed in bold as you speak to get the right pronunciation.

Constellations

Andromeda
AN-**DROM**-EDA

Aquila
AK-WILL-AH

Aries
AIR-RIZ

Boötes
BOH-**OH**-TEZ

Caelum
SEE-LUM

Camelopardalis
KAM-ELL-OH-**PARD**-A-LIS

Canes Venatici
KAN-AYZ VEN-**AT**-I-SEE

Cassiopeia
KASS-EE-OH-**PEE**-AH

Cepheus
SEE-FEE-US

Cetus
SEE-TUS

Circinus
SUR-SIN-US

Coma Berenices
KOH-MAH
BERR-REN-**NICE**-EZ

Corona Australis
KOR-**ROH**-NAH
OST-**TRAH**-LIS

Corona Borealis
KOR-**ROH**-NAH
BOR-REE-**AY**-LIS

Equuleus
EK-**KWOO**-LEE-US

Lacerta
LASS-**SER**-TAH

Lepus
LEP-PUSS

Libra
LEE-BRAH

Monoceros
MON-**NOSS**-ER-OS

Ophiuchus
OFF-EE-**OO**-KUS

Orion
OR-**RY**-ON

Pisces
PY-SEEZ

Piscis Austrinus
PY-SIS **OST**-RIN-IS

Vulpecula
VULL-**PEK**-YOO-LAH

Stars

Achernar
A-**KER**-NAR

Albireo
ALBI-**REE**-OH

Aldebaran
AL-**DEB**-AH-RAN

Alpha Centauri
AL-FA SEN-**TOR**-EE

Antares
AN-**TAR**-EEZ

Betelgeuse
BEE-TELL-GURZ

Dubhe
DOOB-EE

Fomalhaut
FOH-MAL-OH

Iota Orionis
I-**OH**-TA OR-**RY**-ON-IS

Mizar
MY-ZAR

Polaris
POE-**LAR**-IS

Procyon
PRO-**SY**-ON

Rasalgethi
RAH-SAL-**JETH**-EE

Planetariums

A planetarium is a room on the dome of which a projector like the one shown here projects images of stars, planets and other sky objects.

The projector shown is typical of the sort you will see in a good planetarium. Running around the "horizon" of the dome is a silhouetted skyline to increase the illusion of actually watching the sky.

Some major cities have planetariums – among them Toronto and New York.

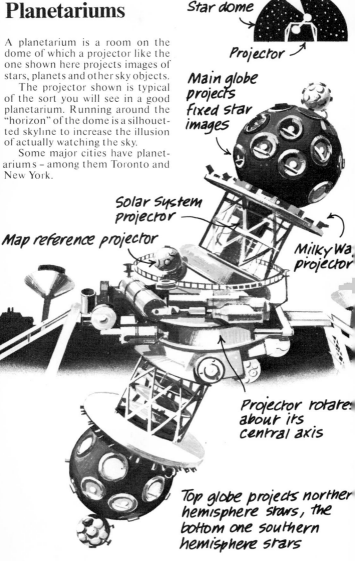

Star dome

Projector

Main globe projects fixed star images

Solar System projector

Map reference projector

Milky Way projector

Projector rotates about its central axis

Top globe projects northern hemisphere stars, the bottom one southern hemisphere stars

Glossary

Here are some definitions of terms used in this book.

Asteroid One of the many lumps of rock orbiting between Mars and Jupiter. A second belt may exist between Jupiter and Saturn. Some asteroids do not stay in these areas – some pass very near the Earth.

Big Bang Name given to the initial explosion of matter which scientists think was the beginning of the Universe.

Binary Two stars orbiting round each other. Usually too close to be seen separately, except with binoculars or telescope.

Eclipse When one object moves into the shadow of another, it is eclipsed. A lunar eclipse occurs when the Moon moves into the Earth's shadow. In an eclipse of the Sun, the Sun is hidden by the Moon for a brief period.

Galaxy A giant group of stars. The Milky Way Galaxy contains about 100,000 million stars.

Hemisphere One half of a sphere, e.g. the northern hemisphere of the Earth.

Light Year Distance travelled by light in a year. The speed of light is just over 186,000 miles a second, so a light year is 6 million million miles.

Local group of galaxies Like stars, galaxies tend to collect in groups. The Milky Way is one of a group of about 30 galaxies.

Meteoroid A piece of rock or a dust particle flying free in space. Called a meteor if it enters the Earth's atmosphere; if it hits the ground, a meteorite. Millions of meteors collide with the atmosphere every year.

Nebula Cloud of dust and gas in which stars are forming or have recently formed.

Orbit Curving path taken by a celestial object when it revolves around another, for example, the Earth around the Sun.

Phases Different shapes that the Moon and planets seem to have as different parts of their surface are lit up by the Sun.

Radiant Central point from which, by the effect of perspective, a shower of meteors seems to come.

Satellite Small celestial object orbiting around another, such as the Moon around the Earth. Artificial satellites orbit the Earth in their hundreds.

Star Luminous globe, powered by the energy of nuclear fusion. Surface temperatures range from 3,000°C – 50,000°C.

Sunspot A cool area on the surface of the Sun. Because it is cooler it looks darker, even though it is actually a scorching 4,000°C.

White dwarf Tiny remains of a once much larger star. Its matter has collapsed to a point at which a spoonful would weigh many tons.

Scorecard

After a night's spotting, write the date next to the objects seen in this scorecard. Keep a record of your nightly scores. The letter N or S next to a score shows that the object is visible only from the northern or southern hemisphere. Other objects, like meteorites, will usually be seen in museums, so score if you see them there.

	Score	Date seen		Score	Date seen
Artificial satellites Flashing satellite	20		Capricornus	15	
Satellite	10		Carina	10S	
Satellite eclipse	15		Cassiopeia	5N	
Aurora Australis	25S		Centaurus	5S	
Aurora Borealis	25N		Cepheus	15N	
Comet	40		Cetus	15	
Constellations Andromeda	10		Chameleon	20S	
Antlia	20S		Circinus	20S	
Apus	15S		Columba	15S	
Aquarius	15		Coma Berenices	20	
Aquila	10		Corona Australis	10S	
Ara	15S		Corona Borealis	10	
Aries	10		Corvus	10	
Auriga	10		Crater	15	
Boötes	5		Crux	5S	
Caelum	20S		Cygnus	5	
Camelopardalis	20N		Delphinus	10	
Cancer	15		Dorado	20S	
Canes Venatici	15N		Draco	10N	
Canis Major	5		Equuleus	20	
Canis Minor	10		Eridanus	10	

	Score	Date seen		Score	Date seen
Fornax	25S		Pegasus	10	
Gemini	5		Perseus	10N	
Grus	15S		Phoenix	15S	
Hercules	15		Pictor	20S	
Horologium	20S		Pisces	20	
Hydra	10		Piscis Austrinus	15	
Hydrus	15S		Puppis	15	
Indus	20S		Pyxis	20S	
Lacerta	20N		Reticulum	15S	
Leo	5		Sagitta	15	
Leo Minor	15N		Sagittarius	5	
Lepus	10		Scorpius	5	
Libra	15		Sculptor	25S	
Lupus	15S		Scutum	20	
Lynx	20N		Serpens Caput & Cauda	15	
Lyra	5		Sextans	25	
Mensa	25S		Taurus	10	
Microscopium	25S		Telescopium	20S	
Monoceros	20		Triangulum	15	
Musca	15S		Triangulum Australe	10S	
Norma	20S		Tucana	15S	
Octans	20S		Ursa Major	5N	
Ophiuchus	10		Ursa Minor	10N	
Orion	5		Vela	10S	
Pavo	15S		Virgo	10	

	Score	Date seen		Score	Date seen
Volans	20**S**		Humorum	15	
Vulpecula	20		Imbrium	10	
Eclipses: Lunar Partial	30		Nectaris	15	
Total	40		Nubium	15	
Solar Partial	35		Serenitatis	10	
Total	50		Tranquillitatis	10	
Meteors Fireball	30		Vaporum	15	
Meteor	10		Oceanus Procellarum	5	
Aerolite	10		Ptolemaeus	20	
Siderite	10		Phases of the Moon First quarter	5	
Eta Aquarids	20		Full	5	
Geminids	20		Last quarter	5	
Lyrids	20		Waning crescent	5	
Orionids	20		Waning gibbous	5	
Perseids	15		Waxing crescent	5	
Quadrantids	20		Waxing gibbous	5	
Taurids	20		Schickard	15	
Moon Archimedes	20		Sinus Iridum	15	
Caucasus	20		Tycho	15	
Copernicus	15		**Nebulae and star clusters**		
Halo	15		Lagoon nebula	20	
Appenines	10		Omega Centauri	10**S**	
Mare Crisium	5		Orion nebula	10	
Foecunditatis	10		Pleiades	10**N**	
Frigoris	15		Praesepe	20	

	Score	Date seen		Score	Date seen
Planets Jupiter	10		Coal Sack	10S	
Mars	10		Large Magellanic Cloud	10S	
Mercury	25		Milky Way	5	
Saturn	15		Mira	15	
Venus	10		Mizar	5N	
Stars and galaxies Alcor	10N		Small Magellanic Cloud	15S	
Algol (varying)	30N		UFO	50	
Andromeda galaxy	20		Vapor trail	15	
Antares	5		Zodiacal light	30	
Betelgeuse	5				

Books to read, clubs to join

How did we find out about Outer
Space? How do we find out about
Black Holes? How did we find out
about Comets? Three books by
Isaac Asimov (Walker).
Colonizing Space. Erik Bergaust
(Putnam).
Quasars, Pulsars and Black Holes in
Space. Melvin Berger (Putnam).
The Nine Planets, rev. ed., and
Black Holes, White Dwarfs, and
Superstars. Franklyn M. Branley
(Cromwell).
Skylab: The Story of Man's First
Station in Space. Wm. J. Cromie
(McKay).
Space Puzzles: Curious Questions
and Answers about the Solar System.
Martin Gardner (Simon and
Schuster).

Our Changing Universe: The New
Astronomy. John Gribbin (Dutton).
Point to the Stars, rev. 2nd ed.
J. M. Joseph and Sarah Lippincott
(McGraw Hill).
On the Path of Venus. Lloyd Motz
(Pantheon).
The Moon: Steppingstone to Outer
Space. Dorothy Shuttlesworth
(Doubleday).

Magazines
Sky and Telescope. (The Sky
Publishing Corp).

Clubs to Join
American Association of Variable
Star Observers.
Amateur Astronomers' Association
(there is a chapter in almost
every state).

Index